"If there is anything from which we can learn, it's absolutely nature;
because nature, it's the biggest and the best circular economy (…)
We belong to (…) nature. And that's why we have to behave like we
are part of (…) nature (…) Environment and economy are two sides
of the same coin."
Janez Potočnik
(New environmentalism and the circular economy: Janez Potočnik at
TEDxFlanders, 04.04.2014)

FOREVERCIRCLELAND

Written by
George Hohbach & Ehrengard Hohbach

Illustrations by
George Hohbach

English translation
George Hohbach with Lynda Brandt

Music & lyrics by
George Hohbach

Arrangement and sheet music by
Alfred Huff

In anticipation of the book, George and Ehrengard planted 100 trees in Tanzania in February of 2021 with the Environmental Charity *One Tree Planted.*

Bibliographical Information of the Deutsche Nationalbibliothek
This publication is listed in the Deutsche Nationalbibliographie of the
Deutsche Nationalbibliothek; detailed bibliographical information
can be accessed under http: //dnb.d-nb.de

Printing and Production: BoD – Books on Demand, Norderstedt

Illustrations and cover design by George Hohbach
My thanks to Mrs. Frick for her support.

ISBN: 978-3-7534-1028-9

Contents

PART 1

FOREVERCIRCLELAND
– p. 1

THE TITLE SONG
– p.45

PART 2

FOR PARENTS & EDUCATORS
THE CIRCULAR ECONOMY (CE)
– p. 48

Once upon a time, when humans on planet Earth still believed that flower fairies, dwarves and giants existed, a young boy called Max lived in a wonderful, magical land called *ForeverCircleLand,* far away from his family and hometown. This beautiful, sparkling land was named in such a way because its inhabitants knew, that in nature everything happened in circles. For instance, water fell on their land as rain, where it collected and then, under the influence of the sun, evaporated back into the atmosphere, becoming clouds in the sky and falling down once more as rain. This was also true for the change of night to day and back again, and for any other circle, including the Circle of Life. Every circle differed in its own way from its predecessor and brought up something new. However, whichever way one looks at it, a circle remains a circle, and this will be forever true.

In *ForeverCircleLand,* Max was studying under the auspices of the Great Magician Libertus Harmony. Performing magic was Max's greatest joy, and, according to his Master, he was a very talented apprentice and student of the magic arts.

To be able to perform magic really well required a comprehensive knowledge of magic words, charms and spells, in order to bring something into existence by Magic Will, and then, to make it go away again or to undo the spell. This is how Max thought of magic. Max knew all the words and charms by heart. What he really enjoyed most though, were the extraordinary magic tricks that he invented himself. On very hot summer days, Max took great pleasure in turning a matchstick into an elephant, who then cooled him off with the fresh water he blew out of his trunk.

The water tasted like strawberries, and when Max was refreshed, he allowed himself to lie in the sun. Everywhere the strawberry water touched the soil, little strawberry plants began to grow, providing large, luscious fruits. By this time, the elephant had changed into a little mouse who began to collect the strawberries and serve them to Max. As the mouse did this, she tickled Max behind his ears with her long whiskers, which made him shake with laughter.

Every time a magic spell was performed correctly in *ForeverCircleLand,* a wonderful sound emerged, sometimes even a melody. Should such a magic sound ever be heard in disharmony, it meant the magic spell was incorrect, and needed to be practiced again and again for as long as it took to make the right sound. This is how the magicians in *ForeverCircleLand* avoided producing things that did not bring joy. Yet it was not always easy to detect wrong sounds or to recognize them at all; to do so required a lot of practice.

Within two years Max had acquired all the knowledge and skills a true magician needed to know, but following an old custom, he took one more year to perfect his talents.

One day he asked his Master if he could now take his Magical Master's Exam, since there was nothing left for him to learn, or so he thought!

"Well, well,'' said Master Libertus Harmony. After looking into Max's eyes for a long time, he spoke: "There are no objections against you taking the exam. You are a brilliant student, perfectly capable of all magic tricks. But I have to warn you, the exam, though it will be most simple, will also not be easy, and those who fail lose all their magic powers."

"Will it be that difficult?" Max asked, as his stomach began to rumble with nerves.

"Of course, it will," was his Master's meaningful reply. "Your Master Exam might be based on, let's say, the task of turning a golden frog into a golden chicken using your magic wisdom."

Max couldn't believe what he had just heard! This was straight forward, one way magic, changing one object straight into another one, like speeding down a one way street. Nothing simpler than that.

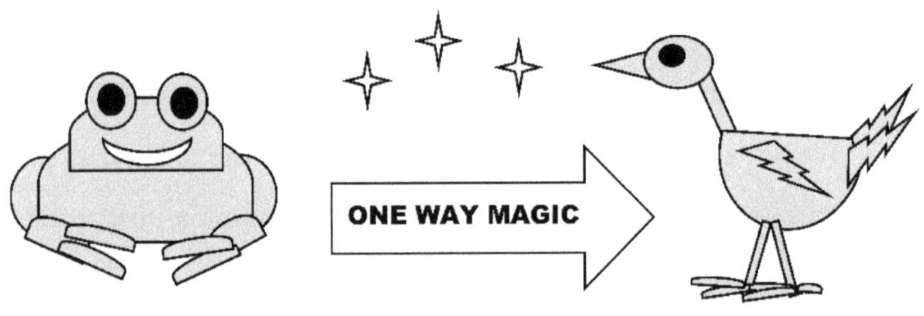

Turning a frog into a chicken was also the easiest magic trick he could think of! For sure, Master Libertus Harmony had been joking when he said that the exam was not going to be easy. So, Max laughed.

"Well then," said Master Libertus Harmony, "are you willing to take this dangerous and challenging exam?"

"Absolutely," Max nodded, feeling cheerful. "No problem. I'm ready any time."

"So be it," replied Master Libertus Harmony. Once again, he looked deeply into Max's eyes. "But before I administer the exam, I must leave to attend the meeting of the Great Magicians which takes place every four years. I will return at the beginning of next month, and you will sit your exam then."

"Great," said Max. "That will give me quite enough time to completely prepare for the exam."

But, secretly in his mind, Max decided he would kill time goofing off until his Master returned, as the exam was just too simple and super easy. Ample time for him to chillax in the hammock in front of the brightly colored magic mansion of his master. Especially as the hammock was attached to the branches of the most gorgeous cherry tree in the middle of the garden.

So as soon as Master Libertus Harmony had left on his flying carpet, Max made himself comfortable in his hammock and dreamt the days away. He pictured himself as a Great Magician performing the most impressive magic wonders. Soon he would be famous and known worldwide.

However, after a while, doing nothing became too boring for Max, and he decided to indulge in his favourite pastime: juggling with his wisdom. Max imagined most of his wisdom as three little colorful balls that he called "wisdom spheres" or "wisdom wonder circles". To make them become visible, he recited one of his magic spells, and then he performed little juggling tricks with these symbols of his wisdom.

But handling just three wisdom spheres also became too boring for Max. Since he had some wisdom left, he decided to create a fourth and a fifth wisdom circle, but when Max began to say the appropriate spell to make his wish come true, a somewhat distorted magic sound came to his ears. Considering how unimportant his spell was, he thought he must be mistaken, and that, in reality, the sound was just as beautiful as always, so he finished saying the magic spell. After that, all his wisdom was transformed into colorful balls. Never before had he converted ALL his wisdom into circles, and never before had he handled five wisdom wonder circles at the same time! It was great fun, and he surprised himself with the way he handled so many circles so gracefully. However, concentrating on so many flying objects soon made Max very dizzy, and his eyes began to shimmer.

He decided he wanted to stop juggling, but he didn't know how. When juggling with three wisdom wonder circles, Max knew in which order he had to catch each single circle in his hands to stop juggling. But when it came to five wisdom circles, his hands weren't big enough.

"What on earth am I supposed to do now?" Max asked himself as he lost control over all the balls at once.

They slipped out of his hands and scattered all over the place, which when performing juggling acts, can happen very easily.

"What a dumb accident!" Max cried.

What had just happened was not at all the way he wanted to solve his problem. Plus, he could sense that his magic wisdom was now also dispersed all over the garden, too far away from him, and that gave him rather a helpless sensation.

Feeling slightly morose, he began walking through the garden to pick his wisdom wonder circles up off the ground. The more he collected, the more he felt his magic power again.

"If only I had already passed my Magical Master Exam, I would now be a Great Magician, and accidents like this would never happen to me."

Max put the wisdom circles in front of him and started counting them. "One, two, three, four…," he counted before realizing the fifth sphere was missing. Max started looking around, but he couldn't spot the missing wisdom sphere anywhere.

"Well, then," he said to himself, "I will just have to undo the spell that created the circles and return all my knowledge back into my mind."

Max attempted to recite the magic spell which was needed to undo performed magic, but soon found he couldn't remember a single phrase. Apparently, it was the missing fifth circle which contained the formula for the required spell of undoing magic.

"Unlucky thirteen!" Max cursed, for now he really had to find the lost wisdom wonder circle.

He combed the entire garden looking for it, but the garden was very big. He looked in every corner and under every leaf, but to no avail. It felt like he was having a nightmare. The wisdom wonder circle was nowhere to be found. Max tried to comfort himself.

"At least I don't need it for my exam, because I only have to change a golden frog into a golden chicken and not undo any magic."

Finally, he stopped searching and went back to the Magicians' house to look after the remaining wisdom circles. However, he realized that he was now faced with another problem. He couldn't convert the wisdom circles he had found back into mere thoughts again, since he was missing the last wisdom circle to undo his magic.

"If I don't find the fifth wisdom circle, I will have to carry my other four wisdom circles with me closely at all times in order to access their wisdom!" Max said angrily.

Annoyed by this, he stepped on an apple lying in front of him and was greeted with a loud cracking sound.

"What was that sound?" Max said to himself, "stepping on an apple never makes a crack like that."

He looked down at his feet. To his surprise and shock, he saw the fifth red wisdom wonder circle which he had just crushed into a thousand little pieces. It was the very circle that contained the wisdom to undo magic.

"I'm so lucky I won't need this one for the exam," Max said, after a brief moment of panic before calming himself down again. "Ah, but it also means, I can't change the remaining four circles back into thoughts for the next three days."

Suddenly he heard extremely unpleasant noises coming from the Magician's house. Laughing and giggling of the utmost misbehaviour. As he turned, Max recognized several little creatures running away as they laughed in every rude way possible whilst passing what looked like little balls to each other. Of course! These little balls were the remaining four wisdom wonder circles, and the little creatures were the evil Anger Dwarves who inhabited the Forest of the Black Souls behind the Magician's house.

"This can't be happening!" yelled a totally bewildered Max. "It can't be true," Max said again, this time quietly under his breath.

He knew all too well that if he wanted to get his wisdom wonder circles back, he would have to follow the Anger Dwarves into the forest right away and ask for their return.

"Uh oh," sighed Max.

He knew that nothing pleased the evil Anger Dwarves more than gaining control over you. It was the greatest fun for them, first to make you mad, and ultimately drive you crazy. It was even said that the dwarves had been known to make people burst with fury and rage.

"If only I hadn't stepped on my fifth wisdom wonder circle…"

As these words faded away, a grim and sullen Max set out for the Forest of the Black Souls. Mocking, scornful laughter, whizzing noises, and sinister whispering greeted him as soon as he entered the forest. Every two steps he glimpsed an Anger Dwarf spying on him from behind a tree, only to hide again once Max caught sight of him.

"I must not lose my patience, I must not lose my patience," Max kept repeating to himself.

At that very moment laughter rang through the forest, and Max looked up into the trees to see he was surrounded by evil Anger Dwarves pointing their fingers at him.

"Look at what we have here! What a shabby, good-for-nothing, lazy bum!" the little Anger Dwarves vigorously mocked.

Max looked down in horror, realizing he was now wearing a grey sack-like outfit instead of his magic suit, and on top of his head was an old, itchy straw hat instead of his magic one. The Anger Dwarves had somehow managed to change his honourable outfit into rags, which, of course, aggravated Max even more. It was against his professional dignity to walk around in such poor clothing! Yet Max didn't react.

At that moment, the tip of his nose started to itch terribly, but as soon as he tried to scratch it, it became longer and longer, so that it was finally impossible for him to even reach the tip of his nose.

"Just look at that, what an ugly long cucumber nose this good-for-nothing has now!" the Anger Dwarves yelled and laughed gloatingly.

Still, instead of reacting, Max stayed totally calm and pretended to be relaxed in order to seem as if he was not disturbed at all. But in reality, he had already started to feel angry at these rude, little evil dwarves.

"It would probably be best to let these Anger Dwarves know right away why I'm here, and that they should give me back my wisdom wonder circles," thought Max. "Maybe they'll be understanding if I tell them how important they are to me."

But when Max started to say this to the Anger Dwarves, all he could utter was: "I'm as thick as a plank."

The triumphant laughter of the Anger Dwarves immediately

echoed through the air. With malicious joy they jumped up and down the trees, did somersaults, and stuck their tongues out at Max.

"He's as thick as a plank! He's as thick as a plank!" they screamed unmercifully at him.

Max instantly realized he was starting to boil. A tremendous fury stemmed from his stomach.

"I won't be able to stand this much longer," Max had to admit to himself. Yet he couldn't even finish the thought because he was suddenly interrupted by a "blatch" as he fell into a ditch which had been covered up by sticks, branches and leaves. The terrible, rotten odor that came from the ditch wasn't mud, but donkey milk. And suddenly Max found himself covered in it right up to his chin.

"Everybody, take a look!" the Anger Dwarves called out. "The good-for-nothing is taking a beauty bath in donkey milk. With his long cucumber nose he probably thinks he's as beautiful as Cleopatra, but really he looks like the ugliest scarecrow!"

The laughter which had begun again, instantly when he had fallen into the ditch, made everything worse. The evil Anger Dwarves jumped down from the trees and gathered around the ditch, laughing at helpless Max. They applauded and nearly died laughing, which made Max desperately mad. He almost started crying, so enraged was he at the little Anger Dwarves. The rage in his stomach grew stronger and stronger, and his head began turning red. Soon his fury would make him burst and explode.

"No, no, fury shall NOT make me burst," Max thought. "I'd rather die from laughing. But how?" he wondered, "how could I die from laughing, when I feel so angry, so very angry that I…"

At that very instant Max began to roar with laughter, which doubled the laughter of the Anger Dwarves. For it is well known that laughter makes you happy. All you need is to be aware of this. So Max laughed and laughed and laughed. He laughed so hard that he started to cry.

All of a sudden, he was totally amused by thinking of his long cucumber nose, and how he had said, "I'm as thick as a plank", and how finally, he had fallen into the ditch like a clown.

The Anger Dwarves gradually fell silent, as never ever before had they experienced a human being who laughed with them about being a subject of one of their pranks. Max almost burst with roaring laughter.

Suddenly he noticed he could speak again, and with more laughing rather than words, he told the puzzled Anger Dwarves that he would like to have a big pillow fight. The baffled dwarves helped Max out of the ditch, and for the rest of the day they had the biggest, wildest, pillow fight one could ever imagine. Finally, all of them lay on the ground totally exhausted.

"We've never had that much fun before," the Anger Dwarves concluded before they all fell asleep, worn out but happy.

The next day the Anger Dwarves thanked Max for all the great fun and invited him to come to the Forest anytime he liked to organize another fun-filled activity for them.

Together, they all said to him: *"The circle of friendship now is our bond. It connects us forever and beyond."*

A happy Max went back home with his four wisdom wonder circles, which his new friends had gladly given back to him.

Back home Max decided to put the four wisdom wonder circles in a wooden chest to keep them safe, promising himself that he would not take them out until the exam. After that, he spent the rest of the day lying in the hammock under the cherry tree in the sun-filled garden enjoying himself.

It wasn't long before Max noticed a Flower Fairy, called Circula, who had come down into the valley to pick wildflowers for her crown. As these fairy elves are the prettiest and most fragile creatures one could ever imagine, Max was enraptured by her appearance. When she smiled at him, Max was totally charmed.

"She seems to like me," Max thought, "otherwise she wouldn't have smiled at me that way, but I can't tell for sure. If she glances at me that way again, then I'll know for certain."

Max carefully snuck a look at Circula and waited for her to

glance at him again, but she did not do so.

"I will have to do something to get her attention." Max pondered the idea and then went inside the Magician's house to the chest where he had stored his wisdom wonder circles.

If he could show Circula, the Flower Fairy, how skilfully he could juggle his wisdom wonder circles, she would definitely be impressed and give him another enthusiastic smile. Although fully aware of his promise not to touch any of his wisdom circles until the exam, Max nevertheless took three of them out of the chest. Only the fourth one, with which he could transform a frog into a chicken, he left behind.

Back outside in the garden, Max promptly began demonstrating his juggling tricks and, indeed, the little Flower Fairy was very impressed. He juggled so fast that it looked like he was actually handling more than three wisdom circles.

The little Flower Fairy enthusiastically followed Max's actions, smiling and applauding after the most difficult juggling stunts. Big red hearts beamed from her eyes, and she was about to fly towards Max as she was so in love with him.

But suddenly the sky turned threateningly dark. Completely startled, the little Flower Fairy looked up towards the sky and immediately flew away.

Before Max could figure out what was happening, he was grabbed from behind by two giant claws and instantly lifted up into the sky. Max realised he had a problem. A Firebird, who was now clutching him very tightly, was flying at lightning speed, but to where?

Max had been shaken so badly during the lift-off, that only now he realized he had dropped his three wisdom wonder circles. They had all disappeared, and were never to be seen again.

Carrying a terrified Max through the sky, the Firebird flew directly towards a huge volcano, out of which enormous masses of hot lava, heaps of ashes and great flames burst.

When Max saw this, he became even more terrified because he knew that inside the mountain lived a dreadful giant called the Fire-

Eater. The Firebird was one of his birds of prey which had been sent out to hunt from the mountain.

The giant was waiting impatiently in front of his cave for the return of his Firebird, and immediately took Max from the bird's beak. As his reward the Firebird received a chunk of meat as big as five hippopotamuses, which he greedily swallowed.

Max was terribly frightened and shaking all over. It was best not to tangle with the giant as far as he was concerned. Inside his huge Fire Cave, the giant put Max onto a table and then sat down next to the table on a very large chair.

The Fire Cave was home to the giant. It was one big hot fireplace inside the mountain, hence the name Fire Cave, where the giant loved to cook his favorite dish: red-hot chunks of stone served with a tasty lava sauce.

Nevertheless, now and then, the giant longed for something different, and Max was very much afraid that the giant might have had him caught for a dessert. He trembled so badly from fear he gradually began to get cold, despite the heat inside the cave.

The giant glowered with big dark eyes at Max, as out of his bowl he slurped from a spoon filled with red-hot stones, and then took a big sip from his cup filled with lava wine. His head started glowing and steam whizzed out of his nostrils and ears, a sight which terrified Max even further. Fear paralysed him, and he felt as if his body was turning into a block of ice.

The giant turned to Max and with his deep voice, grumbled like thunder, "I am very angry with you! You flirted with Circula, the Flower Fairy, and she smiled back at you. I am in love with the elf myself, and I will not tolerate anybody else looking at her! It infuriates me even more that she also smiled back at you. My jealousy drives me completely crazy."

Max was filled with dread as he looked into the giant's eyes, and feared that in his rage, the giant would tear him apart into a thousand pieces. He was so frightened that his nose was already frozen into an icicle.

The giant took another sip of his lava wine, and then growled again: "Every night I dream of Circula, and I beat my head against the wall trying to find ways to make her love me too, but I cannot think of a single way as I am too shy to tell her."

The giant stopped talking and bent down towards Max, even angrier than before.

"And then some student of the magic science and arts comes along, like you, and boasts a little with his wisdom circles, and the Flower Fairy smiles at him! This fills me with jealousy and anger, for I fear I cannot do that magic."

The giant's head had begun to flush dark red with anger as steam continually whizzed out of his ears and nostrils. His eyes turned an ugly black, whereas Max, who was in such shock was breathing snow. Soon he would be completely turned to ice! A miracle was indeed needed quickly to save him from freezing, otherwise he would end his life as a snowman in a fire cave.

"Why don't you say anything?" roared the giant. "You probably think you're too good to talk to somebody like me. But don't think I'm dumb, I, too, know that one and one equals seven."

To prove he was right, the giant showed both his hands to Max. On one hand there were two fingers and, on the other one, five. "You see," repeated the giant, "one and one equals seven."

If Max hadn't almost turned totally stiff by this time, he would have started roaring with laughter, for one hand and another hand makes two hands, and obviously two fingers plus five fingers equal seven fingers. Even though Max hadn't been able to crack a smile because he was so cold, the sudden comedy of the giant's demonstration of his knowledge, had been enough to break the ice that had developed around his body.

"I need to quickly thaw out completely or else I'll have a shivering fit, but how can I do that?" wondered Max.

The main fireplace was too far away, and he couldn't jump into the bowl of red-hot stones as it was far too hot in there, and he'd burn up in seconds. The only possibility to warm up again lay with the terrible giant himself.

Max plucked up all his courage, closed his eyes, and with his last ounce of strength jumped onto the giant's shoulder. At least that was the plan. In the very instant he took off from the table, the giant bent his head down to take a closer look at what the little whippersnapper was about to do, so instead of his shoulder Max ended up landing on the giant's nose. Mmm, how beautifully warm that was!

"Hey, you! What are you doing on my nose?" yelled the giant.

Max suddenly realized where he had actually landed. He stood on the giant's nose and looked directly into his eyes, which were as big as mill wheels. The giant had begun to reach his hand up to push Max off his nose, so Max promptly and very confidently said:

"I saw that you have an eyelash hanging on your eyelid, and I wanted to remove it before it fell into your eye, as I know it would be very bothersome if that happened."

"I do?" the giant asked. He lowered his hand with which he was just about to grab Max. "I didn't realize that."

"Oh sure," Max confirmed, "but don't be afraid. I'll have it fixed in a second. However, it might hurt just a little. "

"Please be very gentle," the giant quietly rumbled.

"But of course," replied Max.

He carefully took one eyelash off of the giant's eyelid and showed it to him. "You see, it didn't hurt at all."

"Thank you," said the giant, very pleased. "No one's ever been so nice to me before."

"Don't mention it," Max said, brushing it off. "By the way, this would be a great way for you to make contact with Circula. You could ask her to help you get the eyelash out, and, for sure, it would start a conversation between the two of you."

"That's a great idea," replied the giant Fire-Eater. "You are a wonderful friend. I'm really so glad I didn't tear you into a thousand pieces a minute ago."

"Me, too," said Max and both of them laughed.

So just as the Anger Dwarves had done, the giant said to him: *"The circle of friendship now is our bond. It connects us forever and beyond."*

The very same afternoon the Firebird, at the giant's behest, flew Max safely home. As he waved goodbye, the giant invited Max to visit him again, saying it would make him really happy.

"Okay," Max said to himself when he stepped back into the

Magicians' mansion, "Master Libertus Harmony will be back in two days, as it will be the beginning of the new month. To be on the safe side and to make sure I don't lose my last wisdom circle, which I'll definitely need for my exam, I'll hand the chest key to the mailman in an envelope. He always delivers mail here every two days, so I'll be sure to have the key back on the day of the exam."

So, patting himself on the back for being so clever, this is exactly what he did.

Max felt very relieved when he saw the mailman leave with the key. From now until the day of his exam, he couldn't possibly touch his last remaining wisdom wonder circle, let alone lose it! Max gave a little twirl and grinned, as he already pictured himself as a Great Magician, which made him feel just so good!

Once again Max went to sit in the garden, and as the afternoon sun warmed his face, he felt extremely content with himself. He had arranged everything perfectly.

Suddenly he jumped out of the chair with a terrible sinking feeling in his stomach.

"Oh no!" he gasped—for he had just remembered this year was a magical twist year in *ForeverCircleLand*, and then the month of February in *ForeverCircleLand* only has 28 days instead of 29. "Today is the 28th of February, and since it is a magical twist year, March will start in one day instead of two. This means my Magical Master's Exam will take place tomorrow, but the mailman with the keys won't be back until the day after that. How on earth am I supposed to open the chest without the key?!"

Max turned as white as chalk.

"Now I've messed everything up," he sighed. "Without my last wisdom circle, I'll never be able to pass the exam."

Max stared into the garden, a miserable look on his face. He was distraught. Heavy rain clouds appeared in the sky, and a rough wind began to blow, but Max remained outside in the garden, dejected.

He was very angry with himself, everything and everyone. Why was life playing such tricks on him? He was trying so hard to do this right! There was nothing else he cared about more than becoming a Great Magician.

"I really couldn't think of a bigger problem right now," grumbled Max.

Immediately, as he said that to himself, the biggest person he had ever seen popped into his mind. It was the locksmith with the name of Firesmith. He was as big and as round as a ball like no one else Max knew. Thinking of Firesmith made Max smile, and as he did, a powerful flash of lightning lit up the sky. For a few seconds the sky was brilliantly lit, and in the far distance Max saw the house of the Firesmith.

"That's it," Max said enthusiastically. "This locksmith is the best of all of them. He will definitely be able to make a new key for me! After all, he made the lock for the chest in the first place."

Immediately Max set out to Firesmith's house, pulling the heavy chest behind him. Firesmith was the nicest and most easy-going person one could ever meet, but he was also very portly and terribly slow when he worked.

Of course, he agreed to help Max, and right away, but as always, he worked very, very slowly. Max sat in a corner of the room and observed Firesmith applying himself to his work. He really took his time.

"Excuse me, my dear Firesmith," Max said politely after quite some time. "I neglected to mention that I'll be needing the key by tomorrow."

"Oh, my goodness!" Firesmith replied. "That means I will have to hurry."

But with the locksmith there was no hurry about anything, so he continued his work with his same easygoing calmness as before.

"If he continues at this pace, there's no way the key is ever going to be finished by tomorrow," thought Max.

Nervously sliding from one side of his chair to the other, Max watched Firesmith. He still hadn't taken the iron out of the fire, and it was already almost midnight, making Max even more nervous. In his anxiety, not only was he sliding vigorously from side to side on his chair, but he couldn't stop playing with his fingers, and his right eye had started to twitch.

Max turned to him: "My dear Firesmith, I really don't want to sound impatient, but we don't have much time left to finish the key."

"Oh dear, but you'll have no sloppy work from me," Firesmith replied kindly. Sighing deeply, he said: "Oh, sweet Lord, how exhausted I am! A little break will do me good."

In a flash, Firesmith had fallen asleep standing up. Due to his weight, he couldn't sit down or else he wouldn't be able to get up again.

"I can't believe this!" Max muttered in despair. "He's actually asleep! He's not going to finish making the key until tomorrow, I can feel it."

Max was now so nervous he could no longer sit on the chair. He started walking around the house, but that didn't help either. In fact, it made him even more nervous, and he began trembling inside. His whole body began to twitch, and every three steps he jumped up into the air, did a somersault or danced a waltz.

"If nothing happens super soon, I'll go crazy," Max mused. "My hair is already standing on end, and somehow it's crackling on my head."

Max did not quite know why his hair was standing on end, but decided the reason for the crackling and sparkling was because he kept walking up and down across the rough stony floor, which had resulted in a charge of static electricity.

"If only I could transfer this energy to the big Firesmith," Max said to himself. "If he were feeling half as excited as I am right now, he would be done with the key in a few hours. But how can I do that?"

Unfortunately, Max couldn't think of any ideas. "I'll just have to wake him up, and tell him that he must continue his work."

Yet no amount of yelling and shouting could wake Firesmith. In great despair, Max finally shook his arm.

In nature, if one electrically charged person touches another person, all the extra energy is transmitted from one to the other. This feels like somebody stepping on your foot very hard.

And that's exactly what happened when Max touched Firesmith, except that the transfer of energy to Firesmith didn't feel like someone stepping on his foot. Due to his tremendous size, it felt like someone tickling him, which caused him to wake with laughter. His whole body was shaking from the tickling prickling sensation. Never before had Firesmith felt so alive and full of energy.

"Let's do it!" Max shouted cheerfully. "Let's beat the clock, Mr. Firesmith!"

Of course, Max didn't have to tell Firesmith twice. Roaring with laughter Firesmith began to work the red-hot iron with his hammer, and before dawn the new key was finished.

Max gave great thanks to Firesmith, who himself was very pleased at his speed in carrying out the work. The two of them agreed that Max should tell the Anger Dwarves to visit Firesmith occasionally and tickle him a lot!

The Firesmith said to Max: *"The circle of friendship now is our bond. It connects us forever and beyond."*

The time had finally come for Max to take his Magical Master Exam. He had been able to return to the Magician's house just in time to unlock the chest, take out his wisdom wonder circle, and hold it tightly in his hand before Master Libertus Harmony flew in on his carpet.

Immediately, both went into the Master's chamber, where the Great Magical Master Exams were always held.

"Are you ready?" Master Libertus Harmony asked Max in a serious voice.

He nodded. To answer verbally was impossible, for he was

much too nervous and excited.

Master Libertus Harmony conjured a golden, brightly shining frog, and put it right in front of Max on the table.

"Ribit," the frog croaked and twinkled at Max in a cheerful manner with his great big dark eyes.

"You know what you are expected to do," Master Libertus Harmony continued, "your exam is based on your ability to demonstrate that you can turn this golden frog into a chicken using your magic wisdom powers."

Master Libertus Harmony stepped back and left whatever was to happen in Max's hands. Max concentrated on the frog. He took off his magic hat, and carefully put it over the frog. In his hand he could feel his wisdom wonder circle. Very gently he began reciting the following spell:

"1,2,3, while the golden frog is hidden

4,5,6, golden frog turns into golden chicken."

Bing, Bang, Puff and Boom, and one instant later a golden chicken sat under Max's magic hat.

"Gagag," the glittering chicken cackled, giving Max and his Master a look of embarrassment when Max lifted his hat.

"Bingo," Max silently cheered. "I've passed the exam."

Then came the acknowledging words of his Master: "Very good, very good, indeed," Master Libertus Harmony commended.

He stepped forward to the table to take a closer look at the shining chicken and the glimmering bowl. Max's face beamed with joy as he glanced at his Master.

He was sure that now his Master would commence his inauguration as a Great Magician, and therefore was very surprised when he heard Master Libertus Harmony say:

"All right then, let's proceed with the second part of the exam."

"The second part of the exam?!" Max was stunned. "W-w-w-what is that supposed to mean?"

He was totally confused, just a second ago he had perfectly turned a golden frog into a golden chicken. He couldn't understand it one bit. Master Libertus Harmony turned to Max.

"The first part of your exam contained the task of turning a frog into a chicken. That was the basis of your exam. Now comes the second part, and that is to show that you can change the chicken back into the frog."

When Max heard these words from the mouth of his Master, a lightning bolt of despair and helplessness struck him. Of course, he had been able to change a frog into a chicken, but to undo it all, no, this he couldn't do anymore. The skill and know-how for the kind of magic required to undo spells had been instantly lost when he had crushed the wisdom wonder circle underfoot in the garden.

It was exactly this wisdom circle, which he had thought would play no role in the Master Exam, that held the skill he was desperately lacking.

Knowing he was now certain to fail the exam, made Max feel very dizzy. Master Libertus Harmony noticed this and looked at him in a very concerned way.

With a heavy heart, Max confessed to his Master, telling him the truth about how he had lost almost all his magic wisdom. Max told Master Libertus everything that had happened to him whilst he had been at the Great Magician's meeting, leaving nothing out, and sadly ending with how he was now unable to complete the second part of the exam.

When Max had finished telling his story, he turned away with a remorseful look in his sad eyes. But at that very same moment, in his mind, he imagined Circula standing in the midst of his friends smiling at him, and then he saw all his new friends smiling too and intently looking at Circula, the great love of his life.

Unexpectedly, the name of the magic land *ForeverCircleLand* darted into his mind too, in huge golden letters. The name formed a rotating circle out of which countless plants and animals emerged, disappeared and re-emerged again.

"That's it!" Max exclaimed to himself. "Nature uses circular motions to be creative, to do its magic! Nature's circular recipe always moves forward in time. Nature never does away with anything, but just keeps transforming. That way it all stays part of the unity of Nature. I can do the same! I don't have to do away with the chicken. All I have to do is to act like Nature and perform magic in a forward circular fashion. That way, I can let the chicken circle forward until it has transformed into the frog again. Nature's magic is not to erase its parts, to undo and go the same straight path backwards in time, but to keep transforming everything and re-using it all in circles. Nature's circular creativity moves forward, just like time does in its cycles."

"Bravo! Bravo! Bravo! Excellent!" Master Libertus Harmony said enthusiastically, as if Max had spoken his thoughts out loud. Master Libertus Harmony continued solemnly:

"Real magicians only and always use the circular recipe of Nature and do everything in circles, as the circle is the most harmonious form and motion. This helps us to always act in balance with Nature's clever, smart creative harmony. So, we never cast anything away, but always make sure things are transformed from one thing into the other via Nature's magical circular motion of creativity. This produces all the marvelous abundance around us. Keeping to Nature's magic recipe also ensures that everywhere in this universe, in all places, and at all times all magicians are on an equal footing and do the same thing. This guarantees that everything stays as one, interacting as a whole unity. On planet Earth, it was the great physicist

Albert Einstein, who revealed scientifically, that in the universe everything is inextricably interconnected and one whole or one unity. Being one unity allows the laws of nature, the rules by which nature operates, to be the same everywhere all the time. This **sameness** that Einstein called **symmetry**, people often refer to as nature's **harmony** or **balance**. All humans can observe this all-unifying harmony everywhere they look, at all places and at all times. All humans are part of this great universe, exhibiting harmony or sameness, deep equality, balance and symmetry."

Max was enthralled by the words of Master Libertus Harmony. This was really what magic was all about he thought. However Master Libertus Harmony had not yet finished.

"The circle best manifests this beautiful harmony, and therefore is the simplest, easiest and most effective way for Nature to be creative. If we magicians do not apply the same magic, that is to say, **Nature's circularity**, our powerful, creative network and everything around us would fall apart and be destroyed."

Max nodded. He now understood how important it was to be a responsible magician by following Nature's guidance. Max said softly and slowly:

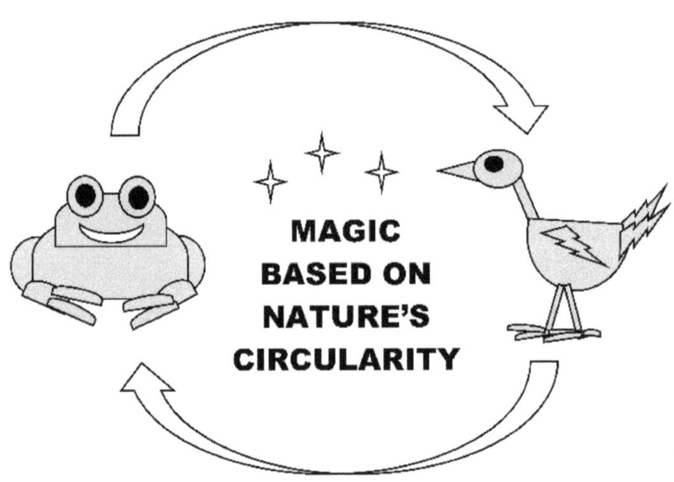

In that very instant, the chicken swirled around in a bright golden light and became the frog again. To Max's surprise he also heard the tinkle of harmonious bells playing a wonderful melody.

Master Libertus Harmony strode up to Max, took his hand and shook it vigorously.

"Congratulations three times! You have passed the Master exam with flying colors, and I am very proud of you. You truly deserve the title of Great Magician, and therefore you too, shall now become part of the circle of Great Magicians."

Libertus Harmony smiled and said: *"The circle of friendship now is our bond. It connects the two of us and all forever and beyond."*

With broad smiles on their faces, they looked into each other's eyes.

At that very moment the mailman burst into the Magician's house. Overcome with excitement he reported that the key Max had entrusted to him had suddenly been transformed into a golden Magic Wand decorated with numerous colorful circles made out of precious diamonds.

Master Libertus Harmony simply beamed even more. After calming the excited mailman and assuring him everything was alright, Master Libertus Harmony took the glowing wand from him.

He handed it ceremoniously to Max, as a symbol of eternal Magic Wisdom based on Nature's magic circular recipe. Max accepted it with great respect and gratitude. He carefully held the Magic Wand close to his ear, and from the wand emerged a very fine tone, tender and delicate. It was the Magic Sound of Nature's cyclic harmony. Master Libertus Harmony smiled contentedly.

"My dear friend, enough deep and profound wise words and noble gestures have been said and done today. Therefore, it is now time for the fun and joyful part of the Great Magical Master's Exam. Come along, Mr. Colleague, let us go now. We mustn't keep the party guests waiting any longer. It's time to celebrate."

And so, they did: Circula the Flower Fairy, the Dwarves, the Giant, the Firesmith, the still puzzled Mailman and, of course, Master Libertus Harmony with his overjoyed colleague Max, who was now a Great Magician. Circula flew towards Max and kissed him. Max looked around cautiously to see where the Giant was, but the Giant, as well as everyone else, nodded and smiled. So Max happily returned the embrace of his true love Circula, and they kissed each other for a long time. *"Love is our bond, forever and beyond."*

When Max, together with Circula, returned to his hometown, from which he had departed several years previously to become a Master Magician, he realized with joy, that his magic had already preceded him. The people in his town had already begun to use Nature's circular recipe in the form of eco-intelligent and climate-smart ways of living. All used materials were recycled, nothing became waste or was discarded. Everything, from the beginning, was undertaken in a way that ensured the consequences of their actions would yield positive effects for everyone—including creatures, both great and small—and create a prosperous present and future.

Max and Circula were welcomed with a great, fun filled festival, celebrating the eco-smart Circular Economy based on Nature's circular ecosystem. The Festival lasted three days and three nights.

Max's friends from *ForeverCircleLand* also joined the fun, heavily disguised as a rock band.

All together they performed wonderful music that made everyone dance joyfully into the small hours.

"The Circle of Nature's Harmony is our bond.

It connects People and Planet forever and beyond."

ForeverCircleLand
Title Song to the Novel

Music & lyrics by George Hohbach
Arrangement by Alfred Huff

★ Also a Youtube Music Video ★

ForeverCircleLand
Song Lyrics

VERSE 1:
Life starts out small
with its recipe of
circularity
to grow tall.

Life sings its song.
Makes reality strong,
circularity
cycles on.

BRIDGE:
Nature is your friend
if you always lend
circularity
a helping hand.

CHORUS:
With circularity,
with Nature's harmony
grows real diversity,
network unity.

See and understand:
magic is your friend
in ForeverCircleLand

See and understand:
magic is your friend
in ForeverCircleLand

VERSE 2:
Life shows its way
with its recipe of
circularity
everyday.

Life sings its song.
Makes reality strong,
circularity
cycles on.

BRIDGE:
Nature is your friend
if you always lend
circularity
a helping hand.

CHORUS:
With circularity,
with Nature's harmony
grows real diversity,
network unity.

See and understand:
magic is your friend
in ForeverCircleLand

See and understand:
magic is your friend
in ForeverCircleLand,
in ForeverCircleLand

PART 2
FOR PARENTS & EDUCATORS
THE CIRCULAR ECONOMY
(CE)

1. <u>THE DEFINITION OF THE CIRCULAR ECONOMY</u>

The Circular Economy is a simple lifestyle based on Nature's circularity. It aims to generate positive, balanced effects for people and planet and a prosperous future for all.

2. <u>THE GOALS OF THE CIRCULAR ECONOMY</u>

<u>The main holistic (all-encompassing) goals:</u>

a. To ensure the wellbeing and the generation of positive effects for people and planet, only so-called *'safe materials'* or substances are used.

b. That the 'safe materials or substances' are of continuously *high-quality* so that they can be <u>fully recycled</u> (and are not just downcycled to end up as waste)

c. The *'safe materials'* are beneficial for people and planet to ensure both the protection and regeneration of the environment: soil, water, air, plant life, wildlife, human beings (this goes beyond just being sustainable, 'keep doing what is being done as long as possible', as the goal is to continuously create positive effects)

Therefore, the positive Circular Economy aims at being eco-intelligent and climate-smart, <u>offering solutions</u> for:

- the (toxic) waste problem (which includes, for example, the

issue of carbon dioxide (CO_2) in the atmosphere and oceans)
- the depletion of resources
- health and environmental issues, in general terms.

These problems are essentially created by the linear economic model of *take-make-waste*.

3. NATURE'S HOLISTIC BLUEPRINT FOR THE CE

The CE respects that Nature is a **BALANCED UNITY-NETWORK**, a <u>positive</u> (all-encompassing) **WHOLE**.

BALANCE comes from the fact that, as physicist Albert Einstein revealed, **SYMMETRY** (harmony, simple sameness) plays the central role in Nature (the cosmos, the universe), because the laws of Nature are **THE SAME** (symmetric, equal, identical) for everybody (all observers) everywhere at all times.

This is so, according to Einstein, as all key phenomena of Nature (the universe), that is, space, time, energy, mass and the constant lightspeed are **ONE** intertwined **UNITY,** or **WHOLE**.

UNITY, in the end, means **SYMMETRY** (harmony, balance, equivalence, sameness) in terms of <u>**doing the same things (having the same effect),**</u> like maintaining the fundamental **UNITY OF NATURE**.

The **CIRCLE** is the simple and most symmetric shape, as one can rotate it infinitely many times *without causing change* to the shape. The shape therefore stays **THE SAME** (symmetric). This respects the aspect of **UNITY** in a very effective and also creative fashion, since the circular motion permits holistic network interactions.

NATURE uses the circular motion on all levels as much as possible in its unity-network on planet **EARTH**. (also see *Why Symmetry runs* *the Positive Circular Economy* by George Hohbach, 2021)

4. <u>THE CE'S SCHOOL OF THOUGHT AND THE BRIGHT MINDS THAT CONTRIBUTED</u>

THE BLUEPRINT

The symmetric

circularity of Nature

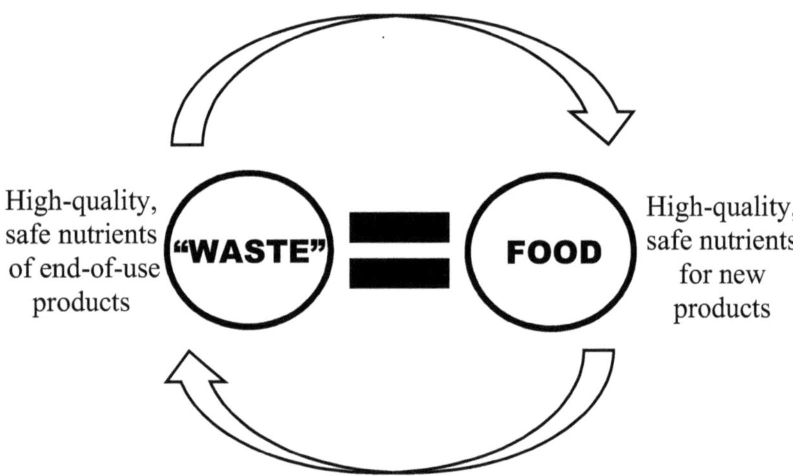

The same (symmetric) circularity applies to both **biological** and **technological** substances (nutrients) according to the school of thought of the holistic design concept of **Cradle to Cradle (C2C)**. In the **biosphere**, the cycles are open, and in the **technosphere**, the cycles are closed loops to keep hazardous but valuable technological nutrients out of the environment.

"…a sustainable future can only be realized if we work together and create new and unusual alliances and partnerships."
Prins Carlos de Bourbon de Parme
(Prins Carlos de Bourbon de Parme: Opening Address,
C2C-Congress, 2017)

The countless environmental problems that humanity is currently facing, are caused by the <u>linear economic model of "take-make-waste"</u>, **<u>that destroys the positive-holistic, circular infrastructure existing in a warped (round) spacetime environment</u>**.

"In all my efforts I have tried to make it clear that all these subjects suffer the same problem because they have become detached from the important basic principles – the principles that produce the active state of balance which is just as vital to the health of the natural world as it is for human society. We call the active but balanced state `harmony´…"
HRH The Prince of Wales
(Harmony, written together with Tony Jupiter & Ian Skelly,
2010, p.5)

"Justice is key to a sustainable economic model if it is supposed to work globally. It's the only way to prevent the ecological question being played off against the social one. Both belong together and can only be solved in unison."
Maja Göpel
(Unsere Welt neu Denken (Rethinking our World), 2020, p.179,
Author translation)

LINEAR is a line and very simply denotes the **MINUS SIGN**, which **is sent through the interconnected network of Nature's POSITVE circles (metabolisms) and destroys them.**

LINEAR/MINUS CHOPPING UP NATURE'S MOST SYMMETRIC, HARMONIOUS, BALANCED, HOLISTIC CIRCLE

MORE

NEGATIVE

On the other hand, the **CIRCLE represents the PLUS SIGN**, that contains both ADDED UP/ADDING UP (UNITY#1) AND THE ADDITIONAL (MORE, #2). So, PLUS can create a CIRCULAR or CYCLIC motion or CURVATURE like this: CONSTANTLY (i.e., ADDED UP as being constant) ADDING UP MORE (THE ADDITIONAL) INTO A UNIT (ADDED UP) TO SHOW MORE TO ADD UP MORE INTO A UNIT (ADDED UP) TO SHOW MORE and so on forever. This creates CURVATURE, CYCLES, CIRCLES and CYCLIC/CIRCULAR MOTION.

Understanding PLUS (ADDED UP/ADDING UP & THE ADDITIONAL) as the circular motion

Constantly ADDING UP MORE THAT SHOWS creates CURVATURE and the Cyclic/Circular motion.

GENERATING THE VIRTUOUS CIRCLE

- **NATURAL NETWORKS +**
- **REGENERATION +**
- **ABUNDANCE +**
- **DIVERSITY +**
- **BEAUTY, HARMONY +**
- **STABILITY +**

MORE

POSITIVE

The linear model is negative, that is MORE by itself. Why? Because, the linear model wants to constantly eliminate (subtract) something, and that something is waste. So, the output is something that is eliminated (subtracted). Accordingly, the concept attempts to destroy UNITY (#1) by subtracting something (the waste) from it. If you destroy unity, then you create MORE in a negative fashion, and since MORE has no end, you manifest INFINITY in a negative fashion. This creation of NEGATIVE INFINITY we can call environmental problems. Take for example the countless toxins or nano-plastic particles now found all over the planet.

NEGATIVITY also means LESS, as it implements the MINUS SIGN. Consequently, we deplete natural resources and cause mass extinction.

The countless circular motions of Nature are being destroyed by linear actions, be it in the soil, water, air or regarding carbon dioxide.

THE SYMMETRIC (POSITIVE) CIRCULAR ECONOMY

The symmetric or positive Circular Economy is a smart, holistic lifestyle, building on the cyclic motion of Nature. It aims towards beneficial effects for people, business and planet.

Thankfully, many people, and the number is growing fast, realize the linear economic model is not working out well. It focusses on being efficient, being MORE in terms of being faster and producing more profit, but excludes the UNITY TERM. Consequently, the linear model is not effective, as it does not implement SYMMETRY (UNITY, BEING POSITIVE, WHOLENESS) on a high, stable level as the positive circular motion does.

> "Nature gave us the correct recipe."
> Michael Braungart and William McDonough
> (The Upcycle, 2013, p.211)

Let's see how the Circular Economy (CE) can be most POSITVE (SYMMETRIC):

The simple, symmetric **Cradle to Cradle (C2C) design concept** co-founded by world-renowned chemist Michael Braungart and pioneering environmental U.S. architect William McDonough (see *Cradle to Cradle* by Michael Braungart & William McDonough, Vintage, 2009), is generally regarded as the guiding path towards the global circular economy from the U.S. to the E.U. to China (see *The Green Industrial Revolution* by Woodrow W. Clark II & Grant Cooke, Butterworth-Heinemann, 2015, p. 368) as well as being the design philosophy at the heart of the circular economy (see *The Circular Economy – A Wealth of Flows* by Ken Webster, Ellen MacArthur Foundation Publishing, 2017, p.16). The WHO regards the Cradle to Cradle design philosophy—with its **cyclic bio- and technosphere** based on Nature's **positive, holistic** concept of the nutrient cycle—as an integral foundation of the CE (see *Circular Economy and Health*, WHO, 2018, p.74). The eco-intelligent and climate-smart Cradle to Cradle design concept **aims at making everything positive from the start for all involved parties. That is to say, people, business and planet.**

> "I've watched as many of the concepts presented in *Cradle to Cradle* have taken root at the U.S. Postal Service and NASA (…) and in countries around the world. I've seen how these simple ideas, when put into practice, can improve productivity and make people happier and healthier."
> U.S. President Bill Clinton
> (Foreword to The Upcycle by William McDonough and Michael Braungart, 2013, p. xvii)

In his profound book *The Circular Economy – A Wealth of Flows*, Ken Webster, circular economy expert and head of innovation at the Ellen MacArthur Foundation, defines the concept of the circular economy as a non-linear system of relationships of parts that work as an all-inclusive dynamic pattern, where all parts interact and influence each other. How we perceive the world, how the economy functions and how we operate within this reality, according to Webster, is a **symmetry**, a matter of sameness, since the natural, circular and holistic premise of the circular economy is omnipresent.

Further, as Webster delineates in chapter 11, this **symmetric state,** in which Nature's wholeness, the systematic interaction of all parts of reality, the dynamics of the circle of this holistic interaction and the collective, organic mindset of humanity are one creative force. It can be achieved globally with the support of information and communication technology (ICT), enabling the principle of access over ownership to become reality. The golden thread of digital interconnectedness in a circular economy (CE) will then be the orchestrated interaction of systematic knowledge, <u>biological and technological materials, (nutrients, the basis of the CE)</u> and clean energy, to benefit both people and planet, and profit in order to prosper long-term while optimizing the entire system.

Given its push towards continuous innovation, the circular economy is both a diverse whole and an open system that evolves based on the convergence of powerful, holistic ideas and scientific discoveries about Nature's wholeness.

"…the solution is all to do with finding a path that puts Nature and her virtuous circles back at the heart of things rather than on the periphery."
HRH The Prince of Wales
(Harmony, written together with Tony Jupiter & Ian Skelly, 2010, p.297)

Consequently, numerous scientists, thought leaders and organizations have contributed and are contributing to the emergence of the positive, holistic circular economy as a school of thought. Among these scientists, pioneers and thought leaders are:

- Albert Einstein, who demonstrated that Nature (the Cosmos) is a whole, a unity, where all aspects of reality work together. The Cosmos is a harmonious, balanced reality. (see *Relativity – The Special & The General Theory* by Albert Einstein, Martino Publishing, 2010). With his work, Einstein revealed the central role of symmetry in Nature, as he not only based his revolutionary discoveries on symmetry as the guiding, primary principle of Nature but also eventually unveiled how central symmetry actually is in Nature.

- Isaac Newton had previously shown that the cosmos is a whole, since on the large scale it is governed by gravity as the uniting phenomenon. Newton's laws of motion conform with each other, working seamlessly in harmony with Nature, again establishing that there is unity, or wholeness in Nature.

- Before Newton, Galileo contributed enormously to the realization of the importance of symmetry (oneness, unity) in

Nature. Albert Einstein consequently achieved his revolutionary scientific discovery about the harmonious unity, simplicity and elegance of Nature building on both Galileo and Newton.

- Physicist Erwin Schrödinger, a contemporary of Einstein, described the actions of a quantum (small particle like an electron) via an overall wavefunction. This overall wavefunction is a holistic whole, consisting of infinitely many single wavefunctions (spreading out over space), each of which has a shape of a <u>revolving</u> corkscrew.

- Mathematician Emmy Noether, confirming the importance of symmetry as unveiled by Albert Einstein also demonstrated that time (which builds <u>on cycles</u>) offers a very flexible "terrain" for symmetry in Nature. (The connection between the continuously symmetric laws of nature in time and energy conservation, i.e., that the amount of energy stays symmetric or the same).

- John T. Lyle with his ideas of *Regenerative Design* establishing the framework of the Circular Economy.

- In his 1966 essay *The Economics of the Coming Spaceship Earth*, economist Kenneth Boulding describes an economic model that is based on nature's circularity.

- World record-breaking sailor, Ellen MacArthur, and Circular Economy Expert, Ken Webster, by promoting the Circular Economy internationally and persuading major players (companies, organizations) to commit to CE practices via the many trendsetting initiatives and partnerships of the Ellen MacArthur Foundation. (see *Full Circle* by Ellen MacArthur,

Penguin Books, 2010; *The Circular Economy* by Ken Webster, The Ellen MacArthur Foundation, 2017; *Sense & Sustainability* by Ken Webster and Craig Johnson, TerraPetra, 2008); Collin Webster, Education Content Manager at the Ellen MacArthur Foundation with, e.g., his TEDxYouth talk: *"Re:Thinking the Future"* or the film *"System Reset"*, about a regenerative economy, on which he collaborated.

"We need to keep the precious materials we have in constant circulation."
Ellen MacArthur
(Full Circle, 2010, p.362)

"... how we see the world; how the economy really works and how we can act within it. There is an important symmetry here."
Ken Webster
(The Circular Economy – A Wealth of Flows, 2017, p.7)

- The Club of Rome with books like *The Limits to Growth* or *Come On!* by Ernst Ulrich von Weizsäcker and Anders Wijkman

"Our greatest inventions, discoveries and acts of creativity come when apparent contradictions are reconciled. Integral thinking is thinking that is able to perceive, organize, reconcile and reunite the component elements and arrive at a truer understanding of the underlying reality."
Ernst Ulrich von Weizsäcker & Anders Wijkman
(Come On!, 2018, p.199)

"The transition to sustainability (…) there will also be changes to (…) the human perspective on nature."
Jorgen Randers
(2052, 2012, p.13)

- Walter Stahel with his concept of the *Performance Economy*, who is also credited with coining the term Cradle to Cradle in the 1970s (see *The Circular Economy – A User's Guide* by Walter Stahel, Routledge, 2019).

"…most of the topics jeopardizing the world society, listed as the objectives of the UN Sustainability Goals (…) could be tackled simultaneously by the shift to a circular industrial economy."
Walter R. Stahel
(The Circular Economy, 2019, p.90)

- HRH The Prince of Wales with his support for organic farming *(*see *Harmony: A New Way of Looking at Our World* by HRH The Prince of Wales, Harper Collins Publishers, 2010)

"… to understand the significance of Nature's processes and to live by her cyclical economy."
HRH The Prince of Wales
(Harmony, written together with Tony Jupiter & Ian Skelly, 2010, p.15)

- Arnold Schwarzenegger, former Governor of California, with his many ground breaking initiatives, such as the Green Chemistry Initiative (see *Total Recall* by Arnold Schwarzenegger with Peter Petre, Simon and Schuster, 2012) and supporting the launch of the non-profit, now called *C2C Products Innovation Institute* in California in 2010. California's EPA (Environmental Protection Agency) recommendation in December of 2008 to:

"MOVE TOWARD A CRADLE-TO-CRADLE ECONOMY."
California EPA
(Press Release, December 16, 2008)

"The time is now for us to go beyond simply being "less bad" and to lead the world in the invention and innovation of *"more good"* with Cradle to Cradle products and a prosperous Cradle to Cradle economy. Together, we will inspire and transform the world."
Arnold Schwarzenegger, May 20, 2010,
(at the launch of the non-profit C2C Products Innovation Institute at Google's headquarters)

- Janine Benyus, establishing the discipline of *Biomimicry* (see *Biomimicry* by Janine M. Benyus, William Morrow,1997)

- Rachel Carson with her seminal book *Silent Spring* denoting how treatment of the environment by humans also impacts themselves.

"The balance of nature (…) cannot safely be ignored any more than the law of gravity (…) it is fluid, ever shifting, in a constant state of adjustment. Man, too, is part of this balance."
Rachel Carson
(Silent Spring, 2002, p.246)

- Gunter Pauli with his concept of the *Blue Economy* that focusses on the holistic big picture. (see *The Blue Economy 3.0* by Gunter Pauli, Xlibris, 2017)

"Nature's principle of optimisation of the whole, rather than maximisation of selected parts (…) involves recognising the delicate balance among all factors..."
Gunter Pauli
(The Blue Economy, 2017, p.7)

- Paul Hawken, Amory Lovins, L. Hunter Lovins, describing the concept of Natural Capitalism (see *Natural Capitalism* by Paul Hawken, Amory Lovins, L. Hunter Lovins, 1999, Little, Brown and Company; also see *Drawdown – The Most*

Comprehensive Plan Ever Proposed To Reverse Global Warming edited by Paul Hawken, Penguin Books, 2017)

- Kate Raworth and her concept of the holistic Doughnut Economics. (*Doughnut Economics* by Kate Raworth, Penguin Books, 2017)

"*Doughnut Economics* sets out an optimistic vision of humanity's common future: a global economy that creates a thriving balance thanks to its distributive and regenerative design."
Kate Raworth
(Doughnut Economics, 2017, p.286)

- Bill Mollison and David Holmgren defining the concept of *Permaculture*, as well as Sepp Holzer.

- Gabe Brown with *Dirt to Soil* about *Regenerative Agriculture* and also providing an excellent definition of holistic management.

"The premise of holistic management is that nature functions in wholes."
Gabe Brown
(Dirt to Soil, 2018, p.36)

- Charles Massy with his book *Call of the Reed Warbler* that focusses on the transition towards an open, holistic mindset in farming.

"So the only way to get our heads right for managing complex creative systems is to think holistically, flexible and openly."
Charles Massy
(Call of the Reed Warbler, 2017, p.338)

- Ibrahim and Helmy Abouleish with the *SEKEM* initiative, pioneering biodynamic farming in Egypt.

- The *Circular Economy Club*, founded by Anna Tari, that globally promotes the implementation of CE principles.

- The *C2C NGO* with its comprehensive program to raise awareness of the C2C school of thought.

- In the article *Impacts of Green New Deal Energy Plans on Grid Stability, Costs, Jobs, Health, and Climate in 143 Countries* (One Earth 1, 449-4463 December 20, 2019), Mark Jacobson et al. examine a scenario where energy comes from Wind, Water and the Sun (WWS).

- The NGO *Kiss the Ground,* promotes regenerative agriculture and provides comprehensive education in this field (also see the book *Kiss the Ground* by Josh Tickell, 2017).

- The foundation *JUSTDIGGIT* works on regreening land to create a positive impact on the climate. Ancient and modern techniques and technologies are combined to regenerate dry land via landscape restoration programs. Educational programs are also offered.

- *Scientists for Future*, a non-partisan and interdisciplinary organization, addresses issues like climate change or crises like the loss of biodiversity and highlights possible solutions.

- Many young climate activists and thought leaders in general have formed organizations and started initiatives around the world to address the pressing global issues and promote solutions, like *Fridays For Future, Re-Earth Initiative, Sunrise Movement, Foreign Policy Youth Collaboration, Climate Live,* etc.

- The *Apparel Impact Institute*, that implements and supports holistic initiatives, led by its president Lewis Perkins.

- In earlier times, polymaths like Leonardo da Vinci, author Johann Wolfgang von Goethe, explorer Alexander von Humboldt, U.S. diplomat and author George Perkins Marsh, U.S. author Henry David Thoreau, the father of the U.S. National Parks, John Muir, British economist Arthur Cecile Pigou, and Charles Darwin, to name but a few, contributed to the holistic understanding of Nature.

- Ancient cultures around the globe also understood both the importance of cycles in Nature and its wholeness as seen in the Yin-Yang symbol, Feng Shui, and Stonehenge.

Apart from the seminal work of former U.S. Vice President Al Gore (see *An Inconvenient Truth*, 2006), it is also of systematic value to consider **the comprehensive findings summarized in the compelling book *Rethinking Humanity* by James Arbib and Tony Seba** (June, 2020). In a profound, tangible and captivating fashion, the authors delineate that the five key sectors <u>information, energy, food/agriculture, transportation and materials</u> are subject to fast, major changes due to sweeping new technologies. This is based on historic developments, the convergence of scientific fields and the fast evolution of technology. Implemented in the correct holistic responsible way, these changes offer opportunities to human society that can both help tackle environmental problems (including climate change) and manifest an unprecedented prosperous future for all in an Age of Freedom.

Various papers, for example, *Completing the Picture – How the Circular Economy Tackles Climate Change* by the Ellen MacArthur Foundation of 2019, *The Circular Economy – A Powerful Force to Climate Mitigation* by Material Economics and its partners, *The Future of Nature and Business Policy Companion* by the World

Economic Forum of 2020, and *Tackling the Climate Crisis and the Corona Pandemic Recession* by the World Future Council of 2020, all emphasize the need to better integrate Nature into our economic concepts. This message is also conveyed by documentaries such as *2040* by Damon Gameau and *Before the Flood* by Hollywood actor/environmentalist Leonardo di Caprio. *Drawdown*, a book edited by Paul Hawken, lists and describes, in a comprehensive fashion, key approaches to overcoming climate change and the environmental crises.

"…to balance the need for growth with the need to maintain stability."
James Arbib & Tony Seba
(Rethinking Humanity, 2020, p.29)

"The growth of business for good incorporates a new meaning of growth which must embrace (…) connecting growth with positive impact on the world."
Marga Hoek
(The Trillion Dollar Shift – Achieving the Sustainability Goals, 2018, p.7)

"An essential element of all living systems is a dynamic balance between self-preservation and integration. Neither aspect is either good or bad. Good is a dynamic balance between both."
Ibrahim Abouleish
(Die Sekem-Symphonie/The Sekem-Symphony, 2016, p.157, Author translation)

"We humans are not stupid. We are not ruining the biosphere and future living conditions for all species because we are evil. We are simply not aware."
Greta Thunberg
(No One is Too Small to Make a Difference, 2019, p.73)

THE SEMINAL REPORT "A SYSTEM CHANGE COMPASS: Implementing the European Green Deal in a Time of Recovery"

"Our actions now will define the future of our society (…) it is no longer enough to simply act. We must act quickly, systematically and together."
Martin R. Stuchtey, Sandrine Dixson-Declève, Janez Potočnik, et al.
(A System Change Compass, October 2020, p.ii)

The seminal systematic report, *A System Change Compass* offers a balanced cybernetic overview, ***a much-needed organizing system***, regarding how a coherent and concise holistic set of key principles (the compass) gives **science-based** orientations on managing socio-economic activities in a fashion that mirrors the holistic way Nature works.

The report was co-authored by members of The Club of Rome and SYSTEMIQ, such as Martin R. Stuchtey, Sandrine Dixson-Declève, Janez Potočnik and others. (see: A System Change Compass, October 2020, p.iii)

The report's symmetric, systematic configuration:
An Overall Organizing System for Human Civilization

The report's symmetric, systematic configuration:
Overall, the report consists of a positive-holistic (all-inclusive), socio-economic and Nature-based system that has **two** linked, interconnected, and equally beneficial parts:

    1) A concise IDEA COMPASS OF CHANGE consists of 10 positive, interconnected principles that define human actions, visions and goals such as financing, governance,

prosperity, competitiveness or consumption. As a whole, it provides holistic empowering orientations or directions for:

2) An equally positive-holistic description of different parts or REAL-WORLD ECONOMIC ECOSYSTEMS of human society, that are kept active and evolve by beneficial, leading, future-fit industries, the so-called CHAMPIONS, interacting in reinforcing positive cycles, in an all-inclusive framework or OVERARCHING SYSTEM OF ENABLING POLICIES.

"It is the predicament of mankind that (…) failure occurs in large part because we continue to examine single items (…) without understanding that the whole is more than the sum of its parts, that change in one element means change in the others."
Donella H. Meadows, Dennis L. Meadows, Jørgen Randers, William W. Behrens III
(The Limits of Growth,1972, p.11)

As a result, the report presents a much needed, holistic (all-encompassing) ORGANIZING SYSTEM for human civilization. It provides a helpful positive guiding system to assist making the necessary economic changes interconnected, mutually beneficial and effective.

THE SYMMETRY CIRCLE
of eco-intelligent and climate-smart concepts and initiatives

The SYMMETRY CIRCLE lists thought-leading concepts that can strengthen symmetry, and Nature's core (balance, network unity, harmony, wholeness, circular action, holism), to make the planet and society MORE SYMMETRIC again. The list is, of course, incomplete and cannot cover all the great, essential efforts. The singled-out, listed key concepts can interact in reinforcing cycles, as they are positively holistic, overlap or complement each other. This way, a more symmetric, hologram-like world view of dynamic unity/wholeness can emerge, be imagined, communicated, felt in its positive, all-unifying force of powerful interacting loops and materialize to benefit all: people, business and planet.

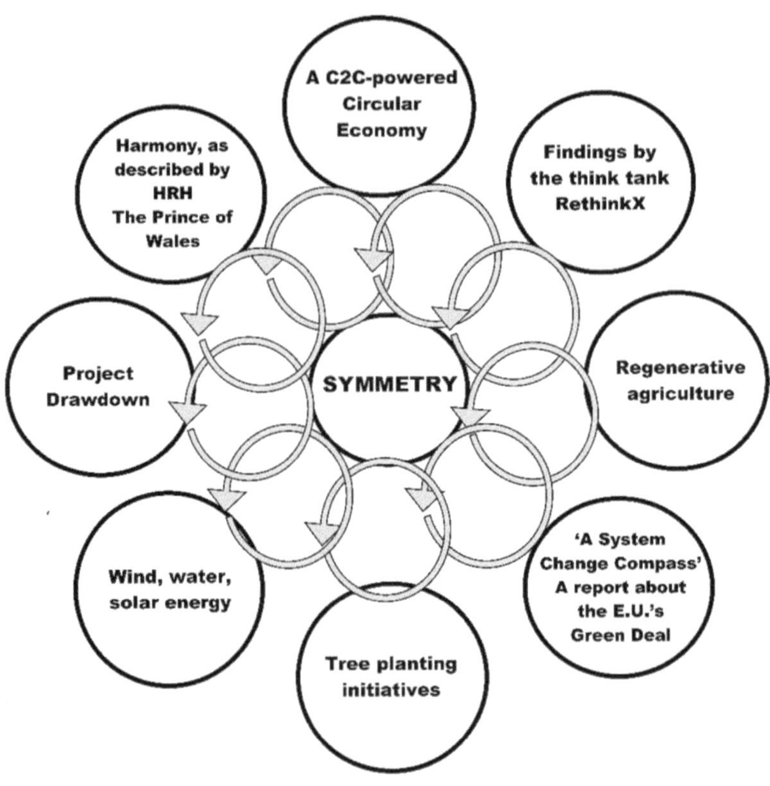

"A System Change Compass: Implementing the European Green Deal in a time of recovery"

<u>THE BIG PICTURE & OVERALL ORGANIZING SYSTEM</u>
The holistic orientation and symmetric, societal Organizing System for governments, companies and society as a whole, as presented by The Club of Rome and SYSTEMIQ.

THE POSITIVE CIRCULAR ECONOMY
Nature's Circular, Symmetric Recipe

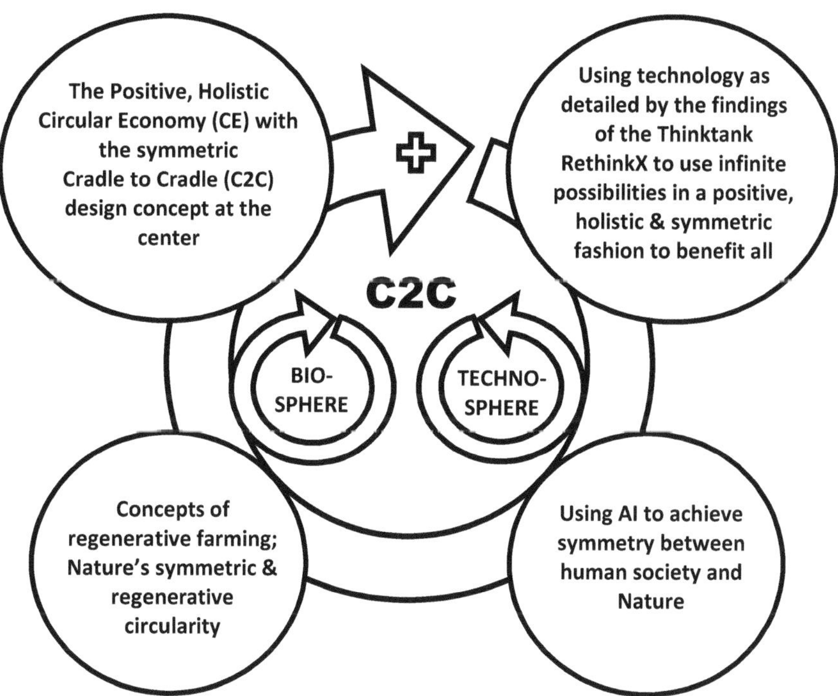

Nature's Foundation:
Albert Einstein's findings on the core role of SYMMETRY in Nature

The implementation of these concepts also supports and promotes the implementation of the UN's 17 Sustainable Developments Goals (SDGs) that present an intertwined whole.

"…if Western Civilization is able to reorganize itself (…) the ability of Western Civilization to survive and go on to increasing prosperity and power will be bright."
Carroll Quigley
(Tragedy and Hope, 1966, p.7)

Will this renewal occur again, also globally, now in the 21st century?

"There is indeed a law, right reason, which is in accordance with nature; existing in all, unchangeable, eternal. Commanding us to do what is right, forbidding us to do what is wrong. It has dominion over good men, but possesses no influence over bad ones. No other law can be substituted for it, no part of it can be taken away, nor can it be abrogated altogether. Neither the people or the senate can absolve from it. It is not one thing at Rome, and another thing at Athens: one thing today, and another thing tomorrow; but it is eternal and immutable for all nations and for all time."
Cicero
(The Republic of Cicero, translation by G.W. Featherstonhaugh, Introduction, p.13)

About the Author George Hohbach

George Hohbach studied Law and has collaborated for nearly two decades with Suite A Management, a talent and literary agency in Beverly Hills. Alongside his writing and involvement with Value Investing, George studied Albert Einstein's revolutionary symmetric equivalence principle for many years. As Head Writer, together with artists from the U.S. and around the globe, he produced several solution-driven Action Comedy Novels aimed at young adults, created around the eco-intelligent, climate-smart school of thought, Cradle to Cradle (C2C), based on the works of its co-founder, chemist Michael Braungart, who developed C2C in cooperation with U.S. pioneering environmental architect William McDonough. Using the connections between Albert Einstein's groundbreaking symmetrical equivalence principle with the ensuing mathematical principles, Cradle to Cradle, and the positive Circular Economy, he has also written and published various scientific summaries on system-theoretical analyses. George has a romantic comedy with U.S. author Robin Palmer about the positive Circular Economy currently in production. He is also working on creative projects inspired by the European report *A System Change Compass*.

George Hohbach is not only a writer but also a composer and artist, and has produced various pop songs on C2C, the Circular Economy or *A System Change Compass*. His prolific artwork in various media, including paintings, music, books, and documentaries, is presented in galleries, educational centers, companies, and museums throughout the world. George lives and works near a beautiful Nature Reserve from where he gathers much of his inspiration.

Ehrengard Hohbach
is a former physician with a specialty in holistic medicine. She gave numerous lectures, seminars and courses on that topic. Her expressionist, symbolic paintings—also inspired by C2C and Einstein's findings about symmetry—are exhibited in galleries, companies and museums domestically and abroad. Her artistic and literary work was covered several times on radio programs as well as in newspaper articles.

Also by George Hohbach & Ehrengard Hohbach with Scot Marcano;
illustrations by Juan Romera:

C2C NOVEL *MICHAEL & MIA*

An Adventure in the Positive Cycle of Nature
illustrated Middle Grade Novel

Two young chemistry fans, haunted by a polluting, mad robot, discover a stunning, circular solution in Nature at the last minute that can save their home planet.

A humorous adventure based on the life of C2C co-founder Michael Braungart with background information on the school of thought of C2C.
Plus: the theme song *NATURE LIVES* as sheet music.

C2C NOVEL *AGENT C2C*

Positive for People and Planet
illustrated Young Adult Novel

Agent C2C and his two American cohorts have to find solutions to two dangerous international crises and the growing environmental challenges – fast!

A humorous action adventure based on the life of C2C co-founder Michael Braungart with background information on the school of thought of C2C. Plus: the title song *AGENT C2C* as sheet music.

Also by George and Ehrengard Hohbach:

THE POSIONITES
& The Global Green Deal
Young Adult Novel

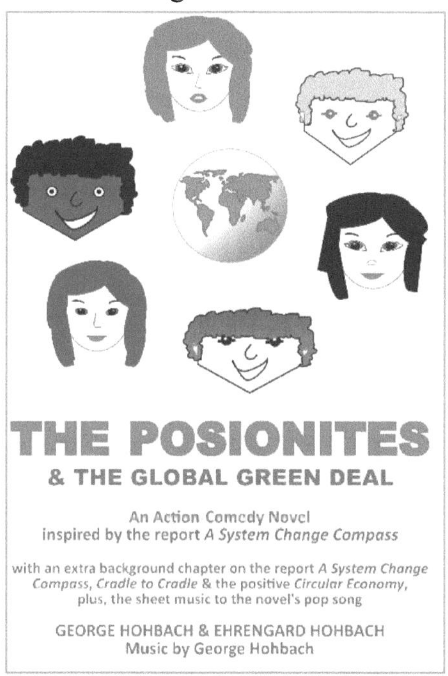

In this Action Comedy, five young science freaks and an actress risk their lives to overcome a dangerous negative force and make a positive future happen based on a pivotal European, eco-intelligent concept.

A humorous action adventure inspired by the European report *A System Change Compass* with background information on the report and holistic concepts like the Circular Economy. Plus: the theme song *SYSTEM CHANGE COMPASS* as sheet music.

MORE NOVELS TO COME...

POP SONGS

& YouTube Videos

Music & lyrics by George Hohbach

ForeverCircleLand
WE LOVE CIRCULAR
SYSTEM CHANGE COMPASS
GLOBAL GREEN DEAL
AGENT C2C
NATURE LIVES
C2C GALAXY
C2C CIRCULAR ECONOMY

MORE POP SONGS TO COME...

**Follow George Hohbach also on
Instagram: @georgehohbach
Twitter: @GHohbach**